INTRODUCTION

I GO BY THE

NAME OF VINCENT RAMOSHABA AND IM 19 YEARS OLD. IM A VERY AMBITIOUS PERSON AND I REALLY LOVE SCIENCE. I BEGAN WRITING THIS BOOK IN 2014 WHEN I WAS VERY INSPIRED BY THE WORK OF ALBERT EINSTEIN AND PAPER WORK OF ISAAC NEWTON. THIS IS MY FIRST BOOK TO WRITE AND I WILL NEVER SAY IT IS MY LAST. I WOULD LIKE PEOPLE TO BE MORE INTERESTED IN SCIENCE BECAUSE MANY THINGS THAT HAPPEN IN OUR UNIVERSE IS THROUGH SCIENE.

Let's come to reality which can bring a lot of sensitive judgement. About approximating the nature of our beliefs concerning the existence of God of which we've never seen how does he behave. We obviously know and had to know that our world is full of not fair diseases, such as diabetes and HIV/AIDS.

When we approach religion with these good thoughts of ours, Christians totally believe that a prayer does cure a disease. According to my view I've never seen a prayer curing HIV/AIDS but I did come across miracles which I know that they exists...but not this time because pastors even do evil magic which God does not allow. The application of chemistry to create significant remedies has solved that problem. Medicine can treat HIV/AIDS and impressively curing a disease like TB. So can we come up or conclude on the idea that science is more powerful than prayer, but it is exactly believed fully by Christians that a prayer is greater powerful to fighting diseases.

Someone told me that whatever you believed at works out for you in a good sense. So if you believed in the sun, it works out for you and if you believe in Jesus he will work out for you. This is A reason why we have different religions.

PAGE 3

So I'm confident that If you believe in science it is more powerful than prayer. I say that prayer is more powerful although I can't prove it and I've never reacted with that.

I remember at the age of 10 after I and my lovely intelligent father had to debate in a promising mind. It wasn't a planned question that I asked him that he doesn't believe that Jesus exists. I had to ask why...and he had no good reason. Because Jesus is the son of God, that means if there's obviously no God the creator, there will be Jesus our lord.

We had no answer to that so I went to sleep in this small shack as it is a good of heat and cold. A question formed in my mind, perhaps it was sent for the future plans concerning the creator .The question went like this...what was there before God. This is a tough question that even a good pastor cannot answer. I stayed with the question for a longer time. Finally an answer came by. My father told me something interesting that if you place a bucket of water in a place where it is not exposed to sunlight a fish would appear from nowhere else.

Page 4

Some people never understand what does this '2016'mean, the year I'm writing these words. We obviously know that 'BC' means before Christ and 'AC', after Christ. What actually means concludes that there's exactly 2016 years after Christ was born, and in a reverse way, about years from year 1 or even a second after and back is not known.

Page 5

C=300 000 000 m/s

THIS IS THE SPEED OF LIGHT

Because God makes anything possible ,that means he can travel at this speed of light .I have tried to think about what will happen when a person moves or travels at this speed of light .That means he can go back in time, lets analyse this analogy …if you move from here to America travelling at this speed , it will take you a second to go there and come back.

So if time and speed of light are equivalent we won't be able to see Jesus coming to this world and going back. Because the time he will spend on earth it will be a second and he will come and go back to heaven within a second. If you don't know yet what is the speed of light , you are lost in physics and brain knowledge.

According to Christian religion, if we say that God made us according to his image that symbolises that he behaves like us. So am I allowed to say that he does excrete, get sick, or even watch television. I just wanted to refresh your minds and I hope that you are enjoying the book.

integrating the sun and the moon

$$\int_{earth}^{space} sun.\,dgod + \int_{earth}^{space} moon.\,dgod$$

THIS WILL BE EQUALS TO THE SOLAR ECLIPSE

The solar eclipse is when the sun and the moon combine and bring a little bit darker atmosphere, Astronomers and astrophysicists are able to predict exactly when the solar eclipse will occur. This shows that the atmosphere obeys and does what the space is expecting it to do.

Scientists have done a lot or even tried to prove that God exists but they've failed. So people who are against them they come up with an excuse to brainwash everyone that scientists are failures because they can't prove that God exists.

Christians have judged and taken scientists for granted but they still use their resources. For example, in church electricity is used...which is the application of electricity. Drums and pianos are being build-up through the techniques of electronics. The predicament is that why can't they appreciate of what they've done to this world because almost everything works through science.

There are scientists who have sacrificed their lives for developing scientific explanations and practical work.

Scientists such as Benjamin Franklin have died in an attempt to study static electricity .Madam Marie Curie died due to more exposure to radiation and was affected by leukaemia. This showed that ancient people were so committed to scientific work. My aim is to prove that God exists by relating it to physics equations. Will I be taking a risk which will lead me to death? But I think God doesn't require anyone to know more about him.

This sounds crazy but I have to split it to you. I really saw God, but I'm not sure of it because I can't prove it. I know that not everyone can see God because it has conditions. The image that you imagine as being God will be him, if you imagine him as being a female...he will be a female. This can be explained by superstition.

Let me just wait a little bit of God. I have accidentally came across the structure of light .But you can only see it when you propagate light into a special different glass.

The history of heaven is what I really wanted to find out about. In heaven there was this angel called Lucifer who was loved and trusted by God. They say he was very good at music, he could raise his wings and make this special sound that could make God jump up and down. So did God know that Lucifer would betray him? If you programme a robot it will do exactly what you want it to do in an exact expected way. God perhaps did not programme Lucifer because if he did, he would not betray him. Did he programme him or did he know that he was going to betray him, we actually don't know. But I know that God really even think of him but we don't know whether he still likes him. This can be explained through this following equation of which I have just formulated myself.

$$\int_{heaven}^{earth} lucifa.d(evil) = satan$$

THE BEHAVIOUR OF JESUS ON EARTH.

Jesus is the son and he has done a very big sacrifice. I have learnt that to change the world you need to sacrifice. An important question is why was Christ white. This is a very powerful question which all of us don't usually think about. Does a second go by where none thinks about this because someone asked me whether does a minute pass where none dies , 'yes' it doesn't. But a second does in this earth.

I am sorry to say this , I really love and respect God but I think he is the one who started racism and even apartheid by making Jesus to be white. Why not dark , think about it ,why white. Perhaps it is because black means darkness and white means good, bright, and peace. Maybe it is true that blacks belong to the dark...SATAN. Whites or non-blacks belong to the lord because he is white even now. Whites have done a lot of wicked things to blacks , perhaps they had a good reason.

But it is true that blacks are not better than whites, this is because most inventions and Nobel prize winners are whites.

The sin of Jesus

Jesus was very strict to his followers and it felt like he disrespected them because he could answer them in anyway. This does make sense but I may be incorrect. The question is...did Jesus sin? I think he did because he drank wine not sure of whether it was alcoholic. In the bible they say that drinking alcohol is not a sin but a sin it is when you get drunk. Obviously if you drink something containing alcohol you're likely to get drunk and Jesus drank one and he got drunk if I think. No this is just to bring sense if I'm wrong in a way that the wine non-alcoholic.

THIS CHAPTER HAS BEEN CLOSED,HOPE YOU ENJOYED IT.

CHAPTER 2;THE ATOM AND LIGHT

Every particle in the universe is made up of smaller particles. An object in this universe mostly in astronomy is called a particle. An atom is so small that a naked eye cannot see it, an analogy can be explained to cancel doubt in our thoughts. An atom can be like the size of a marble at the middle of the stadium when you view it about 2km away from the stadium.

The atom consists of the nucleus, protons, neutrons, and electrons. I neutral. If you propagate light into the nucleus the electrons will gain energy and get excited and jump to the nucleus. To conclude this I can say the nucleus can be neutral when light is propagated into it.

CAN A LIGHT BULB LIGHT UP THE WHOLE UNIVERSE?

This question is not promising, but it can make sense. I have proved that darkness is able to absorb light and light is able to absorb darkness. If we enlighten a room in an isolated system where energy is not lost or or gained, the light bulb can illuminate the room. No matter how big or small the room is.

The universe is in a closed system and one light bulb can enlighten or illuminate the absolute volume irregardless of whether the universe is expanding according to the cosmological redshift.

It has been experimented that if you move an electron at the speed of light it will gain mass. So how does this happen? Actually many have thought that I know everything especially my peers. But this I do not know. I just only think it is similar to when you eat a lot of food you gain a lot of energy that is transformed to overweightness. According to Einstein's equation E=mc (squared) if a body moves at the speed of light squared it gains mass. So an electron if it travels at the speed of light squared it is gaining a lot of energy which is transferred into a greater mass.

As we mostly look up into the speed of light we don't usually look up into the speed of current ,In other words; the speed of electrons. If current moves at 153 miles per hour it is a fact that also electrons move at this speed. This is according to the frame of reference. For example, if you're inside a train that is moving at a certain velocity and you're stationary on the train that means you are moving at the speed of the train.

Light and colour

It is true that there are only seven colours in the universe; red, blue, yellow, orange, violet and indigo. Other colours are a combination of these colours, for example, when you add the primary colours red, blue, and green you get white. I normally know that there's no colour without light. If an apple appears red it has absorbed the other six colours and reflected red.

We usually argue that as us people we don't see, meaning we are blind. This exposes an idea that if there's no light there's no colour and vision. So what do we mean when we say a person is blind? In my opinion it means that the person only sees black or darkness, meaning he or she is blind cause only black is appearing in the eyes. So what if he sees red or any other colour in the spectrum, that means he or she is not blind because a colour is visible in his or her eyes.

Other components of electronics emit colourful colours, we call them 'LEDCs'; light emitting diodes. Let us look at the light bulb that emits white colour, how can electrons flowing bring brightness? This is a simple question to answer, we call it energy conversions. Electrical energy is converted to light energy.

SOMETHING ELSE ABOUT LIGHT

I have been thinking since last year, the year before I wrote this book. I came up with a thought that there's nothing that is except programming. Identical twins will never be exactly the same, even your weight cannot be exact. It is true and not debateable that the speed of light is constant and it think it can't be because its speed is not exact. What will happen if we build up a machine that can travel at the speed of light. Let me make an explanation that can or cannot prove that there's nothing exact besides programming. If the machine and a beam of light take off at the same time will they reach a certain point at the same time. Obviously no because there will be external factors such as gravity and wind that will affect they're speed.

Everything can be affected by gravity, but can light be affected by gravity? Light consists of small particles that are called photons. So do all photons have the same mass or does it depends on the frequency or intensity of light.

LIGHT AGAIN

I promisingly believe that light can produce heat depending on what it is in contact with. There are other light bulbs which has a filament that does not react with argon gas. But as light is produced heat is also produced. This occurs because the exterior of the light bulb is made up of a glass that is a conductor of heat. So light energy can produce heat ,the longer the material is exposed to light it gains heat energy.

Light rays do not have a higher penetrating ability as in gamma rays. This can be proved because there are other objects which do not allow light rays to pass through them. So let's come up with an experiment that explains that light can create heat. If I come up with a good conductor of heat, let's just try to imagine light rays being spread out to that good conductor of heat. If the material has water in it, it will start to boil because heat is being produced.

The boiling point of water is 100 degrees and it's freezing point is 0 degrees. I asked myself a question of what is the melting point of this liquid? I said it is 0 because that's when it starts to melt.

THIS CHAPTER HAS BEEN CLOSED, HOPE YOU ENJOYED IT.

CHAPTER 3; ABOUT THE UNIVERSE

We've been trying to know how the universe came by, but we get answers which we cannot prove whether they are correct or not. According to the dominating religion they believe that God created the universe. But according to science we speak about the 'big bang theory'. So we actually don't know which one is correct but indeed they're both important . We can join up these explanations to cancel doubt. Let's just say that God created the universe using the big bang theory and it was formulated in that way.

Remaining overwhelming thoughts of the theory of evolution by Charles Darwin has brought a lot of misunderstandings. People are over brainwashed because evolution has brought a lot of evidence. Although other people don't understand this, we can say that after the universe was formed evolution occurred to last to bring humans into existence.

Astronomy is very fascinating and it has brought a lot of understanding of how we view and how the universe behaves. We have been debating on whether do aliens exist, not those on earth but those on other planets. I heard on a radio that aliens in space have been seen hugging each other.

We as humans we have been exchanging words not in a formal way concerning the extinction of the universe. It was declared that in 2012 the world will come up into an end. This is because every month there's the 12th day, 12th month, and the 2012th year. So there's no such thing as the 13th month as in 2013. According to astronomy we speak about the 'black hole', a very big empty hole in space.

Astrophysicists believe that when the universe comes into extinction all the planets will go to the black hole and they'll be swallowed.

What if the sun goes to unexistence or disappears. The whole universe will become dark, the question is... what will happen to the planets? This is a good promising question, let us take a look when electricity goes off, all people would bump into each other and cars will collide. Same applies to the planets, they will collide and huge destructions will be made.

Does the sun move? Yes it does. According to what I've seen on youtube they say it moves at about 240km/hr. But I also think that planets revolves around the sun at a constant velocity.

I AND TUMI CONCERNING THE UNIVERSE

I asked this wonderful lady a good question. I said what was going to be there if there was no God. She said there would be nothing, and what is that nothing. The purpose of asking a question is to get an answer and indeed the correct answer. She confidentially said that there is an answer to a question , and the answer is supposed to be correct. I started to know that if there was no God there would be something and that something is nothing.

I said previously that anything you believe at works out for you. If you pray water, the sun, or ancestors, they will work out for you. There is something called faith in this world, if you believe that you will see a ghost you will see it. So if you're really sick just have faith and you will be healed. That's why when you go to a doctor before you take the medication you are already healed. That is all in your mind because your brain controls everything in your body.

That's why when Stephen Hawking started to be paralyzed they told him that he will never walk again...he asked whether his brain would be still be functioning because if your brain is damaged you'll never function.

WHAT ABOUT THIS?

All the components of the universe are contained in one space, the 'universe'. So is there another thing besides that, where were we going to be contained at besides the universe. Perhaps that's why we don't know where God is, he is outer the universe where no-one can reach.

People think it is a fact that when the sun comes closer to the surface of the earth all of us will die. I beg to differ, according to the wave equation, frequency is inversely proportional to the wavelength. So as the light rays of the sun comes closer it's wavelength increases due to compression which causes the frequency to decrease, resulting in a poor penetrating ability.

THIS CHAPTER HAS BEEN CLOSED , HOPE YOU ENJOYED IT.

CHAPTER 4; RELIGION AND SCIENCE

WHAT IS RELIGION?

Religion is what people put their trust on. It began long time ago , others believed in the Egyptian gods and Greeks prayed the sun. Did people believe in Christianity before Jesus appeared? It is true that all religions believe that there's God. There are conflicts between these varying religions, every individual belonging to a different group is desperate to make his religion to be in a dominant state. Although there are wicked wars against this, God gave us a choice to be used.

WHAT IS SCIENCE?

Science is a very big thing that is very vital in our world. The simple definition is 'knowledge'. Science is broad, e.g, astronomy, physics, biology, chemistry, electronics, medicine, etc. The creation of the universe by God is almost and actually it is exactly science. So who is

God, I normally call him a scientist cause he does experiments about his creation and he follows and lives according to the laws of physics. I always attempt to explain this thoroughly but people don't understand what do I mean.

The world is complicated ,others do music, sports, application of technology, experiments concerning medicine, churching, and others even do risky things.

So why are we different? We all have the same structure of the brain depending on the sex state, but we do different things.

THE RELATIONSHIP BETWEEN RELIGION AND SCIENCE.

Science contradicts religion. Science cannot prove that there's or even religion. One of the action that brought power in science is through the cloning of a human being, but it had to bring absolute ends because they were playing God. Religion has brought miracles especially in Christianity. Christians had and will ever believe that everything in the bible that everything in the bible is true. It is written that there will be false prophets in our 21st century and it is happening. I had to say why people believed that Jesus was and is the son of the creator. While I was writing this book, someone told me that Jesus died from a premature death. I didn't believe in Jesus but I believe that he is the son of God and God loves him as he loves every human and any alien in this universe.

PEOPLES DIFFERENT BELIEFS

Jews for Judaism, Christians for Christianity, and Muslims obviously believe in Allah. But there is this religion belonging to Lucifer, so called 'Illuminati'. It almost, but I am doubtful of whether it began in United States of America. It is displayed on the America dollar and it is secretly taking over the world.

CONFLICTS BETWEEN RELIGION.

The moment Jesus became a very important person in the world, people not all but they started to wish to assassinate him. Maybe it was meant to be in that way...to die for us. Muslims and Christians have some conflicts between them. Muslims totally hate Christians and they want them to join their thoughts fully into Allah. Suicidal bombings are done against Jesus' followers, some believe that it is praying by doing that. Between all these beliefs there's obviously this only this only true religion, but it is 'possible' that none is true. So which existing religion is true, the dominating factor maybe. We know that earth is being dominated by the ocean, so let us pray the ocean because majority counts.

WHY PEOPLE HAD TO BELIEVE IN JESUS.

It took a lot of energy for people to believe that Jesus was really the son of God. Ask yourself why people had to believe that Jesus was really the son of God. This is because of the things he used to do, for example, walking on water, saving people from death, doing miracles, and turning water into wine. This really made people to trust in him . It is true that if we had someone who did exactly what Jesus was capable of doing, people would believe that the person is the son of the creator.

It is written in the bible that magic is what God didn't want to see. He called it sinning in a propagandizing way. So can we call what Jesus was doing magic, I don't know but what people are imitating from him is called magic currently. People such as Dynamo, Chris Angel, and prophet Bushiri , they do a lot of amazing things which I may call magic. Are they sinning ? Yes they are. What if Bushiri was the first person to do all these kind of things, and called himself the son of God. Yes people were going to buy that and follow his doings because they were going to be amazed of what he was doing.

HOW WILLPEOPLE REACT WHEN JESUS COMES BACK.

We've been waiting patiently for Jesus to come back. People have been hungry to see how does he look like, and we keep on predicting annually the day he will appear in this world.

One problem we never thought about is how people will react when they see him. Will the respect him, or will they have the gut to crucify him again. I hope that doesn't happen and I hope he will fix all the issues that we have in this wonderful global gatherings. Another problem is what if we are dreaming that he will come and an unfortunate thing happens that he doesn't.

Now I'm about to explain how people react when they see that white dress. Obviously non-Christians won't believe that he is the son of God, even Christians will have doubts concerning on whether he belongs to the one and only. God created us for a reason, perhaps to see his power being in existence. This is like giving birth to a son, you really want to see him doing exactly what you expecting him to do. If he doesn't , you even think of punishing him. So is it true that there's hell, not to mention heaven. God loves us all but according to my view he will take us all to heaven, and if it happens that he is heartless he will definitely take us to hell.

THEORIES AND THEIR SENSE

Lamarck contributed a lot in the theory of evolution. His law of use and disuse has been rejected for a number of reasons. I have come up with an analogy to explain this. Let's imagine a newly born baby who's caged and no-one talks to him they only give him food. Will the baby be able to talk? Obviously the baby won't . The law of inheritance doesn't apply in this real world. Because even God didn't inherit anything from someone else.

Darwin's natural selection is true and simple. If you are born in a very cold place and migrate to a very hot place you won't be able to adapt in an easy way. So the question arrives, will Jesus adapt when he comes back to earth because climate change had occurred. It will take time for him to adapt. Same applies to us, when we go to heaven will we adapt? I hope the bible will answer that.

THE THEORY OF RELATIVITY

I have been trying to understand the theory of relativity and I am not fully satisfied. According to Albert Einstein in 1905 he said that if a train is travelling at the speed of light and you're stationary inside the train you are also travelling at the speed of light. What if you are moving towards the same direction of the train at 50 miles per hour that means you'll be moving at the speed of light plus the 50 miles per hour, absolute motion; 'everything moves relative to each other'.

SUPERSTITION

Superstition is a very powerful topic. It relates to how your mind works. What pops and forms in your mind Is interpreted in the brain. This what I'm about to a say does not sound promising, if you believe that water and fire are not touch them and I promise you won't burn. This is all in your mind. If you believe that you can turn water into wine like as Jesus did, you will. I think magicians also use this sensitive thing called superstition which is related to faith.

HAVE YOU EVER HAD ABOUT THE UFOs?

That is reality. You know what I don't understand about people is that when they don't know the matter, they tend to disagree. A group of American people took a camp, and suddenly they saw a group of flying machines that stood there stationary. When they came closer to the machines they speedily took off. The question is... where did those machines come from? Mostly believe that they came from space and I believe also. Astronomy explanations came this true.

There is something we call 'light years'. This is an Astronomy unit that is misunderstood by many. A light year is the distance that light would travel in one year. It is a fact that there will be the everlasting life, actually there is something that goes on forever. Because the universe keeps on expanding according to cosmological redshift, light travels and follows the not existing boundaries of the universe. So light will go on forever travelling towards these not existing boundaries.

WHY THESE PICTURES?

Someone had to ask me about the difference between an image and a picture. An image can be drawn picture face and a picture gives an exact part of the face. If you take two pictures of Jesus Christ they look exactly the same. So which one is real? The colourful one or the darker one. We cannot be guaranteed about the race of Jesus but we believe that he was Jewish. This means he was non-black, meaning he was white. Why white and not black? This is bothering me and I still have no absolute answer. Whites have been dominating in all and have gone against the smell and dignity of blacks. I will always say that perhaps they have a good reason.

DO ANIMALS HAVE BELIEFS?

Animals such as dolphins,cats ,dogs ,snakes ,lions, e.t.c have cephalisation. They do think, but how and for what reason. Did animals have beliefs? It is a stupid question. The name stupid means talking about something which you don't know. So I don't know whether they do have beliefs. Animals do have something that keeps them entertained, such as dancing and having sex, biologically is called mating.

WHY DID GOD CREATE SEX?

The main reason god created sex is for reproduction. Now people do it for pleasure. It has been written in the bible that sex before marriage it's a sin. That's what the religion Christianity has to say. There are diseases now sexually caused, people have been misusing the truth about the rules of Christianity.

'Adultery' mostly not understood. It actually means a married individual sleeping with other people. 'Last' is when an individual feels like having sex when he or she sees someone opposite of his or her sex. Some girl told me that life is not all about sex. A disease such as bipolar disorder makes a person more sexually active.

THE THEORY OF EXISTENCE

You cannot conclude that something is present without you seeing it. You can only say it is there when you open, find it and see it. What I'm trying to say is that you cannot say there's Vincent interior the room without you seeing him. So something which you don't see and have no evidence concerning it…it doesn't exist. But I do get permission to say that God exists because there's evidence bringing good attitude in our thoughts.

ICONS WILLING TO TAKE OVER

Do you remember Hitler? He has a good story to tell, but it turned out to be wicked. Hitler became a dictator and started his Nazi group that came into power in 1933. He turned Germany into a racist country and every country became to go against it. One more thing I remember about Hitler is that he was about to take over the world and wanted to own the ten commandments desperately. So was he thinking in a correct way or he wanted to take God's position. One more last thing…was he a Christian? In fact he was, a catholic but his deeds turned out to bring wickedness in him.

THE HISTORY OF OUR SOUTH AFRICAN ICON.

Nelson Mandela will be and he is our South African historical icon. He also acted as Jesus Christ and wanted his country to be in a freedom state. He fought for our country politically and with suitable reasons. Before he became the president he studied law at wits university and was the leader of the ANC party. He grew up in a very poor family in Eastern Cape and moved to Johannesburg to look for a job. He then began his career as a law student while he was staying in Alexandra.

He became a hero when he was arrested and sent to Robben Island and stayed there for 27 years. He can out of prison in 1990 and became a president in 1994. All the new rules and laws were established by him.

THANKS TO TATA NELSON MANDELA, IF HE WASN'T THERE I WOULD BE GETTING THIS OPPORTUNITY OF EXPRESSING MY IDEAS IN THIS BOOK.

www.ingramcontent.com/pod-product-compliance
Lightning Source LLC
Chambersburg PA
CBHW030042230526
45472CB00002B/633